Harry C. Gray

Gray's Prescriptionist

Harry C. Gray

Gray's Prescriptionist

ISBN/EAN: 9783337345587

Printed in Europe, USA, Canada, Australia, Japan

Cover: Foto ©berggeist007 / pixelio.de

More available books at **www.hansebooks.com**

GRAY'S

PRESCRIPTIONIST

A TREATISE

ON THE ART OF READING AND COMPOUNDING PHYSICIANS'
PRESCRIPTIONS, WITH TABLES OF WEIGHTS AND
MEASURES, ANTIDOTES, ABBREVIATIONS, ETC.

By H. C. GRAY, Ph. G.

Author of "Gray's Pharmaceutical Quiz Compend," "Gray's Elements of
Pharmacy," and "Gray's Clinical Urinalysis."

CHICAGO, ILL.:
GRAY & BRYAN, PUBLISHERS, BOX 593,
1891.

PREFACE.

The compounding of physicians' prescriptions constitutes one of the most important duties of the pharmacist, requiring more skill and experience than any other branch of his business.

I have endeavored in this little work to present to the reader such practical hints and directions on the subject as I have found useful during many years at the prescription case, together with such tables and other information as may be found useful for reference.

THE AUTHOR.

THE PRESCRIPTION.

A prescription (Latin *præ*, before, and *scribere*, to write) consists of five parts.

The Superscription,—consisting of a single sign, ℞, an abbreviation for recipe, meaning take thou. In French prescriptions the letter P., or the word Prenez, meaning take, is used for the superscription.

The Inscription,—or body of the prescription containing the names and quantities of the drugs.

The Subscription,—or directions to the dispenser.

The Signature,—or directions to the patient or nurse, usually headed by Signa or simply S.; and lastly the prescriber's name and the date.

The names of the ingredients themselves are written in the genitive case. Where a liquid preparation is to be made the words Mistura fiat. are used, meaning make a mixture, or simply Misce, "mix thou." When powders or pills are to be made Misce et fiant pulveres, or pilulæ is written, meaning, Mix and let there be made powders or pills. Or for pills may be written, Fiat massa et dividenda (make a mass and divide). The following is also much used, especially by German physicians,—Misce et fiant tales doses numero X, which means "Mix and make ten such doses." The constituents of what may be called a classical prescription are generally arranged under four heads, viz:

The Basis,—or active drug proper.

The Adjuvant,—or substance intended to assist and hasten the action of the basis.

The Corrective,—to limit or otherwise modify the same; and,

The Excipient,—or vehicle to bring the whole into a pleasant and convenient form for administration.

A prescription may, however, contain the basis alone, or the basis with the adjuvant, or the basis with a simple vehicle or diluent. A single ingredient may serve a double or treble office, as in the case of some compound syrup or tincture. Again the basis may need no aid in doing its work, or corrective of its action, nor any special vehicle for its administration. On the other hand there is no limit to the number of ingredients which may be used, provided there is something to be accomplished by each, and also provided there is no chemical or physiological incompatibility between them. The following is an example of a theoretical prescription:

Superscription—℞.

Inscription.—
{ Morphinæ Sulphatis *Grs. iv* Basis.
 Tincturæ Aconiti... 3 *i* Adjuvant.
 Syrupus Rhei...... ℥ *ii* Corrective
 Aquæ Anisi Q. S... ℥ *iv* Vehicle.

Subscription.—Misce secundum artem.

Signature.--A teaspoonful every three hours.

9-29-91.   T. B. Gray, M. D.

The above prescription was written for facial neuralgia. The adjuvant (tr. aconite) assists the basis, morphine, by decreasing heart action. The syrup of rhubarb acts as the corrective by overcoming the constipating effect of the basis, and the anise water serves as the vehicle in which the rest are administered.

PRESCRIPTION WRITING.

The first thing necessary is that the physician be provided with suitable writing materials, pencil and paper, or a fountain pen will be found much better than a pencil, as the writing of the latter is liable to be blurred and defaced by rough handling of the prescription.

It is most advisable that he provide himself with proper prescription blanks giving his name, residence and office hours, also space for a number, date and the patient's name. If he prefers writing his prescriptions in the metric system, the blanks should by all means have the decimal line. The foregoing precautions are of great assistance.,

1st. In assisting the patient or pharmacist in locating the writer.

2d. Encourages putting the patient's name upon the prescription, which so often prevents mistakes in delivering as well as in administering the medicine.

3d. By having good, roomy blanks the necessity for crowding and abbreviating are obviated.

4th. The decimal line takes the place of the

decimal point, which is so liable to be misplaced or omitted altogether, as often happens, or of an accidental dot or fly speck on the paper altering the quantity.

5th. By providing spaces for a number and date, you greatly assist the pharmacist in filing away and retaining the prescription for future reference or re-filling if desired.

In writing prescriptions the physician should take particular pains to write plainly, that the pharmacist may have no trouble in understanding his wishes and directions as expressed therein. This is one of the most important matters connected with the writing of prescriptions, but unfortunately physicians as a rule do not seem to realize that upon the legibility of the prescription depends to the greatest extent its chances of being properly compounded. Each article should be designated by its full Latin name and in the genitive case, unless only a certain number of an ingredient is to be specified, when it should be in the accusative. Whenever abbreviations *are* used great care should be taken to make them as full as possible and write especially plain. If it is desired that the prescription should not be refilled it should be so stated on the prescription, as it is customary with pharmacists to refill them as often as requested unless otherwise directed by the physician. When an unusually large dose of any particular drug is written for, the quantity should be underlined thus, Strychninæ Sulphatis *Gr., i* or some other mark attached to show the pharmacist that such was really intended. This saves much anxiety to the careful pharmacist and occasionally saves delay occasioned by seeking the writer to see if an overdose has not been ordered.

Whenever possible the physician should make the patient understand that the prescription is merely his directions to the pharmacist to prepare certain medicine and is then to remain in the hands of the pharmacist (or words to this effect might be printed upon the prescription blank). This not only leaves the prescription where it belongs for future reference and mutual protection of writer and dispenser, but would to a great extent settle the long discussed question as to whom the prescription belongs, and prevent its being hawked about among the friends of the patient, doing incalculable damage by being used in cases for which it was never intended. A copy of the prescription could be furnished the patient upon the order of the physician, where it is necessary, as in the case of transient patients or travelers.

In writing extempore prescriptions it is a good plan to first write patient's name and then the names of each ingredient, after this is done you decide on the number of doses to be ordered and calculate the quantity of each ingredient from that number. When the same amount of two consecutive substances is ordered the quantity is omitted after the first one, and aa is placed after the second. This means that the quantity following it applies to both of the preceding ingredients. We might say in conclusion that physicians should aim to render their mixtures as pleasant to the taste, smell and sight as possible, for these details go a great way toward the success of practitioners.

RECEIVING THE PRESCRIPTION.

Upon receiving a prescription from a customer the first duty of the pharmacist is to see that the patient's name is attached to it. Many pharmacists upon receiving a prescription give the customer a numbered check, a duplicate of which is attached to the prescription and subsequently to the package after the medicine has been prepared, and serves to identify it when delivered to the customer. The writer does not consider this a very good method, knowing of several cases in which it has failed in its purpose through the checks being left on the counter by the customer or being otherwise exchanged. A much better plan is to see that the patient's name is upon the prescription and when the medicine is prepared place the name upon the outside wrapper as well as upon the label; this serves to identify it thoroughly when delivered. Some pharmacists advocate the use of prescription cases, which leave the prescription clerk and his manipulations in full view of the customers with the view, no doubt, of making an impression upon the public by his neatness and dexterity. But there are many serious objections to this, first of which is that customers could not be kept from questioning and otherwise interrupting him. Then if a prescription happens to require a little thoughtful consideration before beginning manipulations the customer is very liable to jump to the conclusion that you are unable to read it or that there is something wrong with the prescription. Many other objections could be offered, but we consider that the two mentioned above are sufficient.

READING PRESCRIPTIONS.

After receiving the prescription from the customer and seeing that the patient's name is attached to it the pharmacist should proceed to his prescription case at once. I would strongly advise against the perusal of the prescription in the presence of the customer, as this is quite liable to provoke such questions as (what is in it ? or, what is that prescription for ?). These are questions a pharmacist should avoid the necessity of answering by all means possible, as the pharmacist has no right to tell the customer the character or medicinal effect of the medicines called for by the prescription. If, however, such questions cannot be parried it is best to state frankly that professional etiquette forbids you to discuss the subject. The next step is to read very carefully the entire prescription to see that you are perfectly familiar with all its details, at the same time satisfying yourself that there is no overdose of any ingredient, and deciding upon the order in which the ingredients are to be mixed. Very little can be written to assist the beginner in learning to read prescriptions as his ability to master this branch of the profession depends almost entirely upon his general knowledge of drugs and his actual experience in handling physicians' prescriptions. Therefore, the beginner should make it a constant practice to read and thoroughly master every prescription filled in the store each day; this can be done during his hours for study, and is about the only way to gain proficiency.

In reading a prescription, if the name of an ingredient is somewhat obscure or doubtful, one is often assisted in determining it by referring to the quantity ordered, the other ingredients or the form in which the medicine is to be prepared; either of these will sometimes suggest what the obscure item is. though they more often have no bearing on the matter, If it is the quantity of an ingredient you are in doubt of, it can often be decided by considering the dose of the ingredient. As for an example where one is at a loss to decide whether a character is intended for a dram or an ounce sign.

Whenever you are not perfectly familiar with any article in the prescription refer freely to your dispensatory or other works of reference. The pharmacist who considers himself so competent that this is never necessary is a dangerous man, and young men too often fail to avail themselves of this help through false modesty or fear of being thought incompetent. We might add that while one should never be ashamed to refer to his books for guidance, it is not always advisable to do so in the presence of your customers.

Omissions.--Very often physicians omit to specify the quantity of one or more of the ingredients in a prescription or the number of powders, pills etc., into which it is to be divided. In such cases there is only one course to pursue, viz., see the doctor and have the omissions rectified.

Overdoses.—When a doubt arises as to an apparent overdose of some dangerous drug in a prescription, the pharmacist must use good judgment and quick decision, bearing in mind that his first duty is to protect the patient and next to protect himself and the physician, and that a physician's mistake does not excuse a pharmacist before the law. The first point is to gain time without exciting suspicion in the mind of the customer. This can be done by informing him that it will be some time before the medicine is prepared, and offer to deliver it or request him to call for it at a specified time. After thus disposing of the customer see the writer of the prescription and satisfy yourself as to whether or not it should be dispensed as written. If the physician cannot be found at the time and you consider it necessary to dispense the medicine, either for the good of the patient or to protect the physician, do so by reducing the dose of the dangerous ingredient to a safe limit and notify the physician at the earliest possible moment, and any honorable physician will appreciate your precaution. It sometimes happens that even after such cases have been brought to the notice of the doctor he will insist that the dose is all right. In such cases if the pharmacist still thinks it an overdose he should refuse to dispense it and if he has reason to consider the doctor ignorant and incompetent it becomes his duty to inform the customer his reasons for such refusal, shough a great deal of precaution and judgment thould be exercised in this matter.

The pharmacist, having familiarized himself with the contents of the prescription, should next select the bottle, box or other container, and if it be a bottle fit it with a proper cork. This should always be done before filling the bottle that the cork may not be soiled if the first one tried should not fit, and that particles from an imperfect one may not drop into the medicine from which it is sometimes difficult to remove them. Next write the label, always putting the patient's name and the date upon it; write plainly and neatly, avoiding all flourishing.

COMPOUNDING PRESCRIPTIONS.

We are now ready for the actual operations of compounding the medicine, a point on which very little can be said in general, but for which your whole previous education and training were intended to prepare you. In all of these operations should be observed the most scrupulous neatness and accuracy. Each bottle after being used, should be set aside in the order they were used, and after the medicine has been prepared they can again be compared with the prescription, to see that no error has been made. Some pharmacists follow the system of double checking, viz.—having another clerk take the prescription and let the dispenser repeat from memory the ingredients and quantities used. Either of these methods is good, but the former is more practical, from the fact that one clerk is so often alone in the store that it is better to follow a system in which he depends upon himself entirely., The physician often leaves the selection of some ingredient, such as a solvent or excipient, to the judgment of the dispenser. In such cases the pharmacist should be careful to select such ones as will not retard or in any way interfere with the action of the other ingredients, and make a note on the prescription of the article and quantity used for guidance in refilling. It will also be found of great service to number the ingredients on the margin of the prescription in the order that they are mixed, that the medicine when renewed may not differ in appearance from the original.

INCOMPATIBILITY.

Incompatibility is divided into three classes, viz: Therapeutical, Chemical and Pharmaceutical.

Substances are said to be therapeutically incompatible when their action on the human system are mutually antagonistic. With this class however the pharmacist has very little to do.

Chemical incompatibility, is where two substances brought into contact, react on each other in such a way as to form one or more entirely different compounds. Many prescriptions, however, which are therapeutically correct would come under this head; for example, lime water and calomel or sulphate of zinc and acetate of lead. These are prescriptions in which the chemical reaction between the ingredients is understood and intended, but what the pharmacist must guard against are those chemical reactions between the ingredients of prescriptions which form new and poisonous compounds, and have been overlooked by the physician.

Pharmaceutical incompatibility is where two or more substances when mixed react on each in such a way as to throw each other out of solution, causing precipitates or a disagreeable looking or tasting mixture.

The incompatibility of drugs is governed in the majority of cases by a few simple rules which we will give below.

1st. In mixing any salt with strong acids, decomposition is very apt to take place.

2d. Alkalies should never be mixed with salts of the metals proper. Decomposition takes place, and their bases are precipitated.

3d. Vegetable astringents precipitate albumen, gelatine, vegetable alkalies, and numerous metallic oxides, and with salts of iron produce inky solutions.

4th. Glucosides should not be mixed with free acids.

5th. Double decomposition will not occur between solutions of two salts, unless, by the interchange of the two baselous radicals, a substance will be produced which is either insoluble or volatile.

6th. When a solution of a compound is brought in contact with a solution of another compound, and, by an interchange of radicals, an insoluble compound will be rendered **possible,** that compound **will** be precipitated.

Below we give a list of the volatiles, and soluble and insoluble salts, when these are thoroughly learned one, can readily see where double decomposition will occur.

Volatiles.—The volatile substances are H_2O—CO_2—H_2S Hcy—H I—H Br—HCL—HNO_3—NH_3 and HNO_3.

Insoluble Salts.—All Hydrates, Carbonates, Phosphates, Oxides Sulphides, Arsenates, Arsenites, Borates Tanates and Silicates, except those of Sodium, Potassium and Ammonium, are insoluble.

Soluble Salts.—All the compounds of Sodium, potassium and ammonium. All Nitrates, Acetates, Chlorates, Permanganates, Lactates and Hyphosphites are soluble.

All Bromides, Chlorides and Iodides, except those of Mercury, Silver and Lead (the $Hg Cl_2$ is, however, soluble). All Sulphates, except those of Barium, Calcium and Lead.

Incompatibles.—Comp. infusion of cinchona with comp. infusion gentian.

Essential oils with aqueous liquids in quantities exceeding one drop to one fluid ounce.

Fixed oils and copaiba, with aqueous liquids, except with excipients.

Spirit of nitric ether with strong mucilages.

Infusions, generally, with metallic salts.

Compound infusion of gentian with infusion of wild cherry.

Tinctures made with strong alcohol with those made with weak alcohol.

Tinctures made with strong alcohol, with infusions and aqueous liquids.

Discretion should be used in mixing the following with other chemical substances, as decomposition is likely to occur:

Acidum Hydrocyanicum,   Tr. Ferri Chloridi,
Acidum Nitro Muriaticum,   Tr. Iodinii,
Liquor Hydrarg. et arsen.,   Potassii Cyanidum,
    Iod.,   Potassii Permanganas,
Liquor Patassii Arsenitis,   Potassii Iodidum,
Liquor Calcis,   Ferri Citratis,
Liquor Iodinii Compositus,   Zinci Acetas
Liquor Potassæ,   Potassii Acetas,
Liquor Morphia Sulphatis.

SUBSTANCE.	INCOMPATIBLE WITH
Acacia	Alcoholic and ethereal tinctures; Iron Chloride; Lead salts.
Acids, in general	Alkaline solutions; Metallic Oxides.
Salicylic...	Iron compounds; Patassium; Iodide; Lime water.
Tannic	Alkalies, Carbonates; Lime water; Chlorine water; Albumen.
Bismuth Subnitrate.	Calomel; Sulphur; Tannin.
Chloral Hydrate...	Alkalies, Carbonates; Ammonium, and Mercury Compounds
Iodine	Ammonia; Alkalies; Carbonates; Chloral; Metallic salts; Starch.

Lead Acetate ...	Acacia; Acid Hydrochlor.; Acid Sulphuric and Sulphates; Ammon. Chloride; Carbonates; Lime water; Iodine; Patassium Iodide; Tannin.
Mercury Bichloride.	Patassium Iodide; Salts; Carbonates; Tannin.
Mild Chloride (Calomel):	Acids, Acid Salts; Alkalies; Carbonates; Ammon. chloride; Iodine; Potassium Iodide; Iron Chloride, Iodide; Sulphur.
Potassium Chlorate...	Calomel; Organic substances; Sulphur.
Iodide	Lead and Mercury salts; Potassium Chlorate; Silver Nitrate; Chlorine water.
Permanganate	Ammonia, salts; Alcohol, Glycerin; Ethereal oils; Organic substances.
Bromide...	Acids, minerals; Chlorine water; Mercury compounds.
Silver Nitrate	Acids, Acetic, Hydrochloric, Hydrocyanic, Sulphuric, Tartaric, and their salts; Alkalies; Carbonates; Iodine; Potass. Iodide; Bromide; Sulphur.

SYMBOLS OR SIGNS USED IN PRESCRIPTIONS.

M. *Minimum*, the 60th part of a fluidrachm.

Gtt. *Guttæ*, drops.

Gr. *Granum*, or *Grana.* A grain, or grains. The $\frac{1}{480}$ part of the Troy ounce, the $\frac{1}{5760}$ part of the Troy pound, or the $\frac{1}{7000}$ part of the avoirdupois pound.

℈. *Scrupulus* vel *Scrupulum.* A scruple, equal to 20 grains.

Ʒ. *Drachma*, a drachm, equal to three scruples, or 60 grains.

℥. *Uncia*, an ounce Troy; or, in liquids, the 16th part of a wine pint, or the 20th part of the imperial pint.

℔. *Libra*, a pound Troy weight.

O. *Octarius*, a pint.

Fl. *Fluid.* Used as a prefix to certain measures to distinguish them from weights; thus *fl* ℥, *fluiduncia* ; and *fl* Ʒ, *fluidrachma.*

Ss. *Semis*, half. Used as an affix to weights and measures ; as ℥ ss., *semiuncia ;* Ʒ ss., *semidrachma;* ℈ ss., *semiscrupulum.*

NUMERALS.

The following Latin numerals are employed in prescriptions :

I.	 1	X.	 10
II.	 2	XV.	 15
III.	 3	XX.	 20
IV.	 4	XXV.	 25
V.	 5	XL	 40
VI.	 6	L.	 50
VII.	 7	XC.	 90
VIII.	 8	C.	100
IX.	 9		

METRIC PRESCRIPTIONS.

There is much discussion at the present time regarding the adoption of the metric system, and though we do not think it will ever come into general use in this country, it is necessary for the pharmacist to be acquainted with the methods employed in writing metric prescriptions, as they are occasionally met with, and require unusual care in compounding, from the fact that the writers as a rule have a very limited knowledge of the system, and a much less knowledge of doses when expressed therein. There are two methods employed in metric prescriptions—the Gravimetric and Volumetric.

In the Gravimetric both solids and liquids are weighed. This method is in use exclusively in Germany, but seldom met with in this country.

In Volumetric prescriptions the gramme and its subdivisions are used for weighing solids and the cubic centimetre (sometimes called fluigramme), for measuring liquids.

The following show the different forms in which metric prescriptions are written:

R.

    Pulvis Opii...................1 |
    Plumbi Acetates.............. | 5
    Saponis....................1 | 5

M.—Fiant pilulæ XV.

R.

    Quininæ Sulphatis............ 2.
    Strychninæ.................. .015
    Acid Sul. Ar................. 1.5
    Syrupi q. s. ft...............120.

M.—Sig. Teaspoonful after meals.

R.

    Morphinæ Sulphatis.......... .1 *Gm.*
    Bismuthi Subnitratis......... 1. *Gm.*
    Aquæ Anisi ....20. *Cc.*
    Syrupi q. s. ft...............32. *Cc.*

M.—Sig. Shake well and take a teaspoonful as directed.

COMPOUNDING LIQUID PREPARATIONS.

In compounding a prescription the first thing to be considered is the order in which the ingredients should be mixed. This requires a thorough knowledge of the characteristics of each ingredient, and if the dispenser is supplied with this, the rest is comparatively easy. Very often a chemical reaction between two of the ingredients is to be avoided as far as possible. This can often be done by mixing each of them in weak solution with mucilage of gum arabic or thick syrup before mixing them together. On the other hand, the reaction may be desired and must be facilitated by the dispenser by bringing the two ingredients directly together. The addition of a liquid solvent is generally necessary to start the reaction. In such cases the two ingredients should be mixed in a mortar with a solvent and the reaction facilitated by trituration. The product of this operation should never be put into the bottle until the reaction has entirely ceased, as it is liable to explode or force out the stopper. Again, the ingredients of a prescription may not mix at all, unless put together in the proper order, as in the case of compound emulsions, where it is necessary to mix the oil and gum with a part of the water before adding the other ingredients. In nearly all prescriptions the final appearance of the mixture depends upon the order in which the ingredients were put together. Of course there are some prescriptions written that require no special order in compounding, but one must always be on the lookout for those that do, or he will find that he has a dark, cloudy-looking mixture, where a little care exercised in the order of mixing would have given a fine, clear one. Do not mix strong alcoholic preparations directly with aqueous ones, but rather mix with the weaker alcoholic preparations first, so as to reduce them gradually.

When a salt is to be dissolved in a liquid, always put it into a mortar and make a solution before putting it into the bottle. This not only hastens and assures complete solution, but prevents small particles of dust, etc., so often found in the salts, from getting into the bottle. A mixture should under no circumstances be allowed to leave the prescription case with a powder or crystals in the bottom of the bottle, if it is possible to dissolve it by thorough and prolonged trituration.

Filtering.—One is often puzzled to decide whether or not to filter a solution. This should be done only when an excess of some ingredient has been ordered, the removal of which will in no way interfere with the medicinal properties and action of the medicine.

Additions to Prescriptions.—The pharmacist is often tempted to improve on the doctor's directions by supplying some harmless solvent or a more convenient excipient, in order to secure a finer preparation or to facilitate the compounding. This is an unsafe procedure, however, and should never be done without first securing his consent. Of course the above does not apply to cases where it is abs-a lutely necessary to use a solvent in order to make a permanent solution. The following prescription illustrate some of the difficulties of reading and compounding physicians' prescriptions :

Fig. 1.

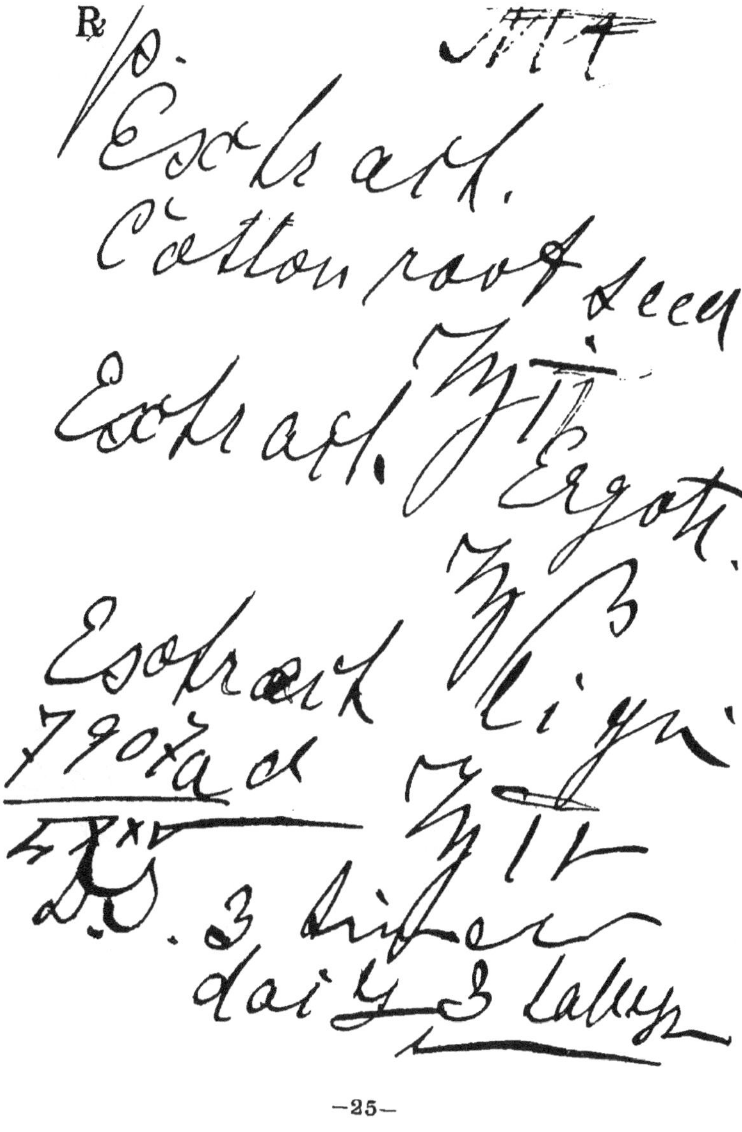

Prescription No. 1 was written by a very intelligent physician, but he seems to have been suffering from absent mindedness when he wrote for cotton root seed, and signified the dose as three tablespoonfuls three times a day. However the pharmacist will have no trouble in interpreting his intentions as follows: Extract of cotton root bark, two oz.; extract of ergot, half an oz, and sufficient fluid extract of licorice to make four oz. The directions should be three times daily a tablespoonful. The error in the directions is the more serious, from the fact that the pharmacist could not always see that such was not intended.

FIG. 2.

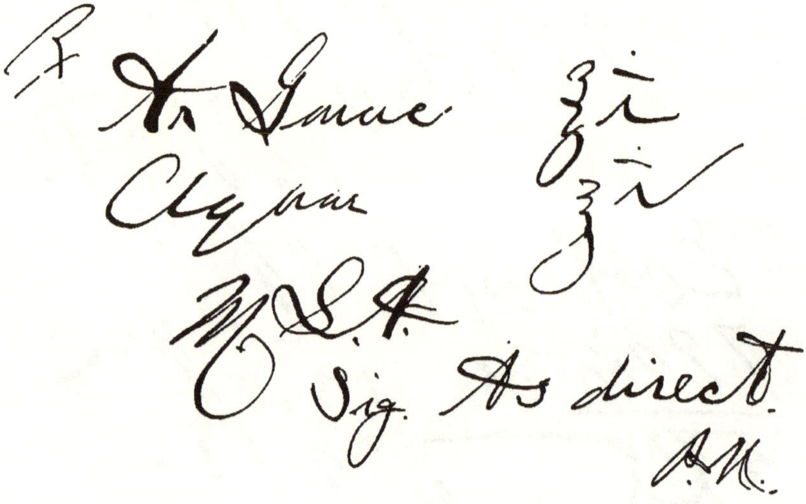

Prescription No. 2 is one that requires particular manipulation in compounding, in order to secure a thorough admixture of the ingredients. The tincture of guaiac should be placed in a dry mortar and by adding the water, a few drops at a time, with constant trituration, a perfect solution is obtained.

FIG. 3.

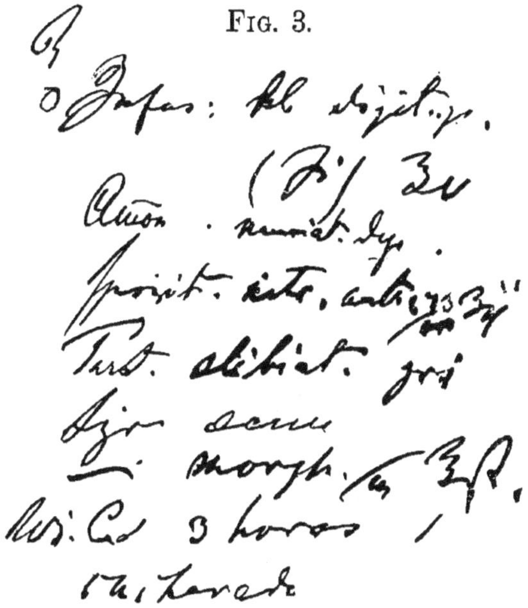

Prescription No. 3 shows very poor writing and abbreviating; it is generally read as follows:

R. Infusi (foliorum?) digitalis......... ℥ v
Ammonii chloridi
Spiritus ætheris nitrosi.............aa ℨ ij
Antimonii tartratis...............gr. j
Syrupi morphinæ (?).....aa q. s. ad. ℥ vj
Misce et fiat solutio.
Sig.—One teaspoonful every three hours.

However, many pharmacists contend that the fifth ingredient is syrup of squills, instead of syrup of senna. Unfortunately, we were unable to consult the writer.

.Fig. 4.

Prescription No. 4 is easily read, in spite of the poor writing, but the pharmacist will see at a glance that the prescription is incompatible.

The sodium arseniate would be converted into potassium arseniate with the simultaneous formation of sodium carbonate. A second portion of the potassium carbonate abstracts from the strychnine salt its sulphuric acid, leaving the much less soluble strychnine alkaloid. Finally the mercuric bichloride is decomposed by the potassium carbonate to give rise to mercuric oxide.

The medicine when prepared would be a six ounce mixture, containing potassium arseniate, pure strychnine alkaloid, oxide of mercury, and a reduced amount of potassium carbonate.

As this obviously is not the doctor's intention, he should be consulted and shown the incompatibility of the prescription.

Fig. 5.

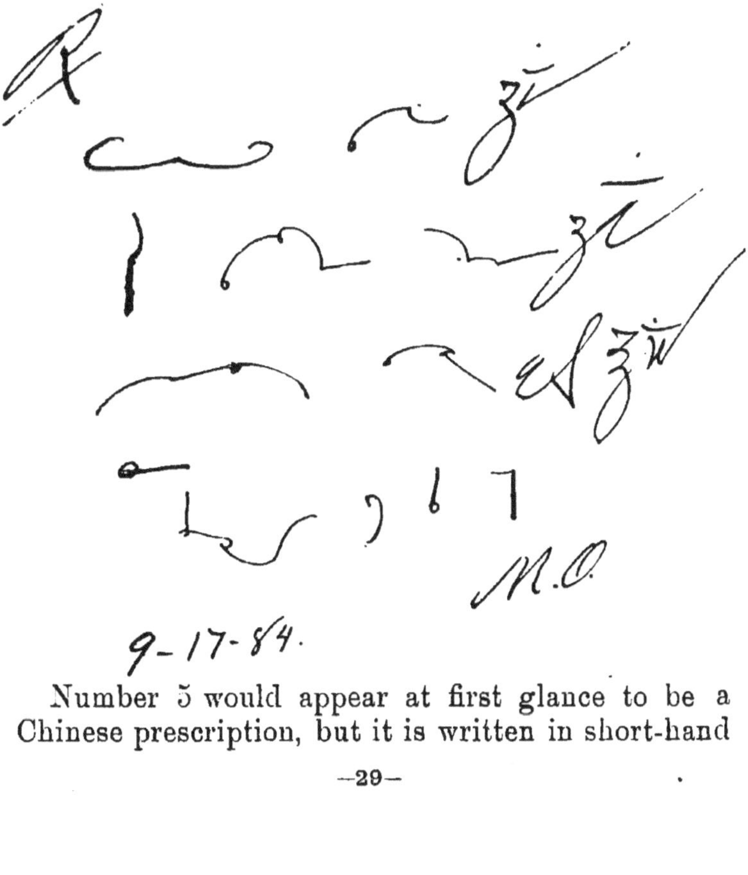

9-17-84.

Number 5 would appear at first glance to be a Chinese prescription, but it is written in short-hand for the purpose of compelling the patient to take it to a certain pharmacist who has added that system to his many accomplishments. Such prescriptions are rarely met with, and the pharmacist could not be expected to read them. This one calls for one dram each of sulphate of quinine and aromatic sulphuric acid and sufficient simple elixir to make four ounces. The directions are a teaspoonful three times a day.

FIG. 6.

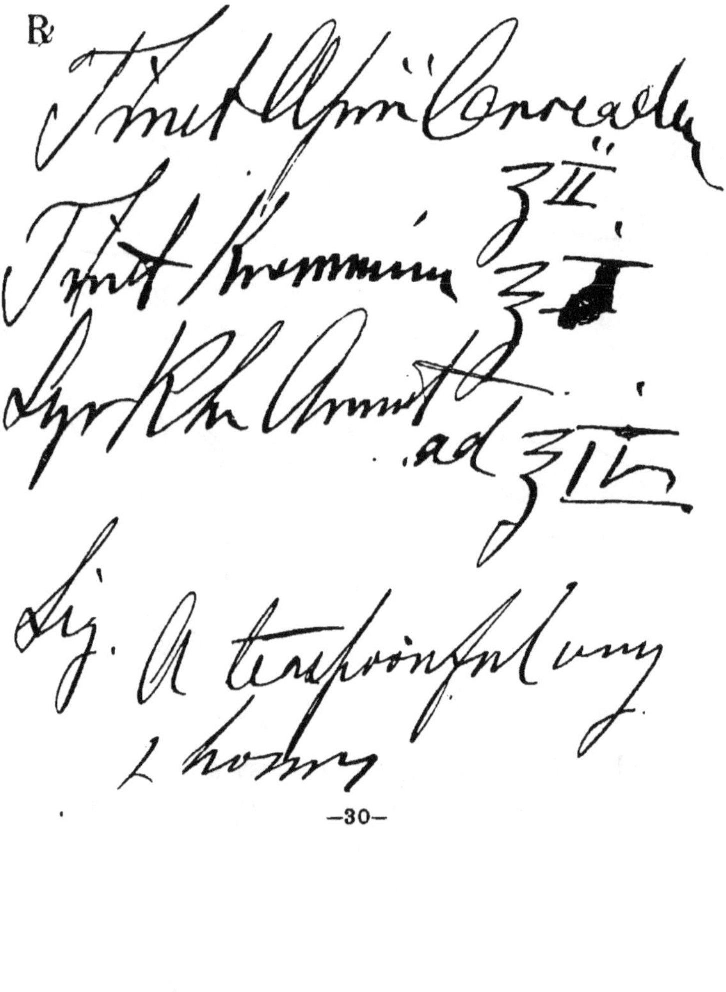

Prescription No. 6 proves a stumbling block to many from the fact that the first ingredient Tinct. Opii Crocada is from the German Pharmacopœia, and seldom used in this country. It is a 12 per cent. tincture of opium.

FIG. 7.

Prescription No. 7 is another very poorly written one and calls for three ounces of Iodia and directs it to be given, a teaspoonful before meals.

The following prescription is one in which the order of mixing is important:

R. Potass. Carb.............four drachms
    Syrup Tolu ....one   "
    Oleum Amygd.............two   "
    Aqua....................three ounces

The carbonate of potash should be dissolved in about two ounces of water, the syrup and oil added and well shaken, then add the rest of the water.

R. Tinc. Benz. Co............two drachms
    Liq. Morph................one ounce
    Mucil. Acaciæ...........half an ounce
    Aqua....................two drachms

Now in this case, if you add water to the tincture, the benzoin is all precipitated and rises to the surface, and it is impossible to mix it, but just shake the tincture well with the mucilage, then add the water, and you get a nice mixture.

FIG. 8.

Prescription No. 8 shows a flourishing style of penmanship which is met with quite frequently and renders it almost impossible to comprehend the writer's intentions. This one is translated as follows: Syr. Hypophos. Comp. (Fellows), seven ounces; Tr. gentian compound, one ounce; Misce. Sig. Teaspoonful three times a day. The short abbreviations together with the flourishing makes it almost impossible to read it until one knows the contents.

Fig. 9.

Prescription No. 9 was written by the same physician as No. 8 and shows that he could write plainly if he would. This one is easily read as follows: Liquor potas. arsen., one ounce; Sig.—five drops three times a day. One might be at a loss to decide whether one or two ounces is called for in the above prescription. But considering the small dose one is prepossessed in favor of the opinion that the second vertical line is an elongation of the one dot above the horizontal line. Such proved to be the case.

FIG. 10.

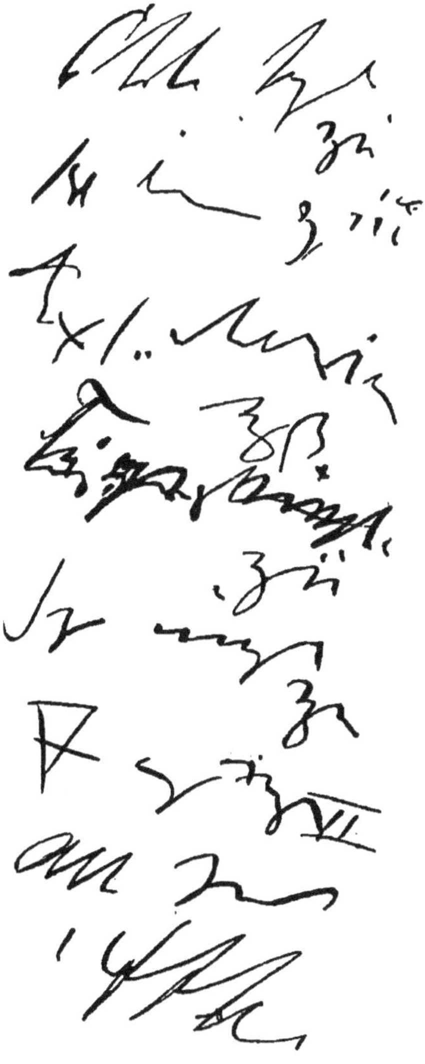

Prescription No. 10 is so poorly written that it is impossible to decide with any degree of certainty what some of the ingredients are, and as the writer is deceased they will remain uncertainties. Almost every pharmacist to whom it has been submitted differ in their interpretation. The following is the most general translation:

R   Chloralis.................two drachms
      Potassii bromidi...........three   "
      Extracti valerianæ fluidi....half an ounce
      Tincture gentianæ..........two drachms
      Syrupi Simp..............one ounce
      Aquæ. q. s. ad............six ounces
      Misce et fiat solutio.
Signa—One tablespoonful at night.

No pharmacist would be justified in attempting to fill such a prescription without consulting the doctor, for his interpretation could be nothing more than guess work. If the writer can not be seen the best plan is to refuse to fill the prescription.

FIG. 11.

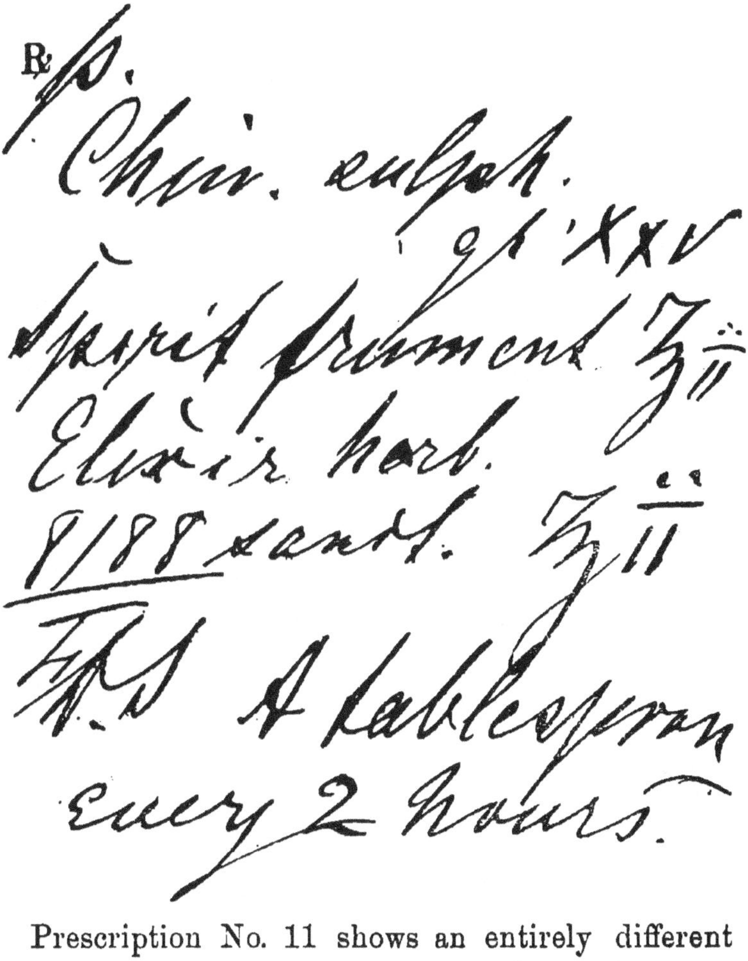

Prescription No. 11 shows an entirely different
style of penmanship and is very easily read.

R.

Tr. Ferri chlor...........four drachms.
Mucil. acaciæ..............two ounces.
Acidi Acet.................one drachm.
Glycerine................four drachms,
Liquor Ammon. Acet. Q. S. add eight ounces.

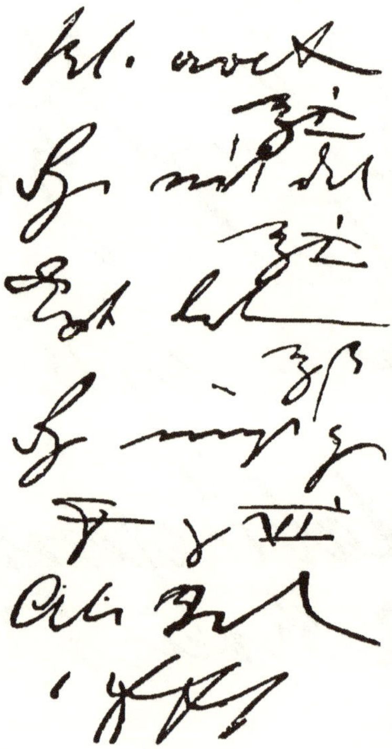

If the ingredients of the above prescription are mixed in the order they are written, a gelatinous precipitate will be formed which would require some time to redissolve. By mixing the tincture of iron with the glycerine and acetic acid, and adding this to the liquor ammonia acetatis, and lastly the mucilage, a nice clear solution is obtained at once.

Prescription No. 12 is another in which some of the ingredients can

FIG. 12.

only be surmised. We will leave its translation to the reader.

INFUSIONS.

Infusions are made by subjecting the drug in a crushed condition to the action of hot water in a tightly closed vessel for about two hours and then straining. The drug should not be boiled in the water. When the strength of an infusion is not directed by the physician or specified in the pharmacopœia it should be made ten per cent. However, the strength of infusions of all powerful substances should be specified by the physician when ordered in his prescription.

Prescription No. 13 calls for half an ounce of rhubarb root, one ounce of senna leaves, with sufficient boiling water to make three ounces of infusion in which one ounce of rochelle salts and two drachms of extract of licorice are to be dissolved. The directions are one teaspoonful every three hours.

FIG. 13.

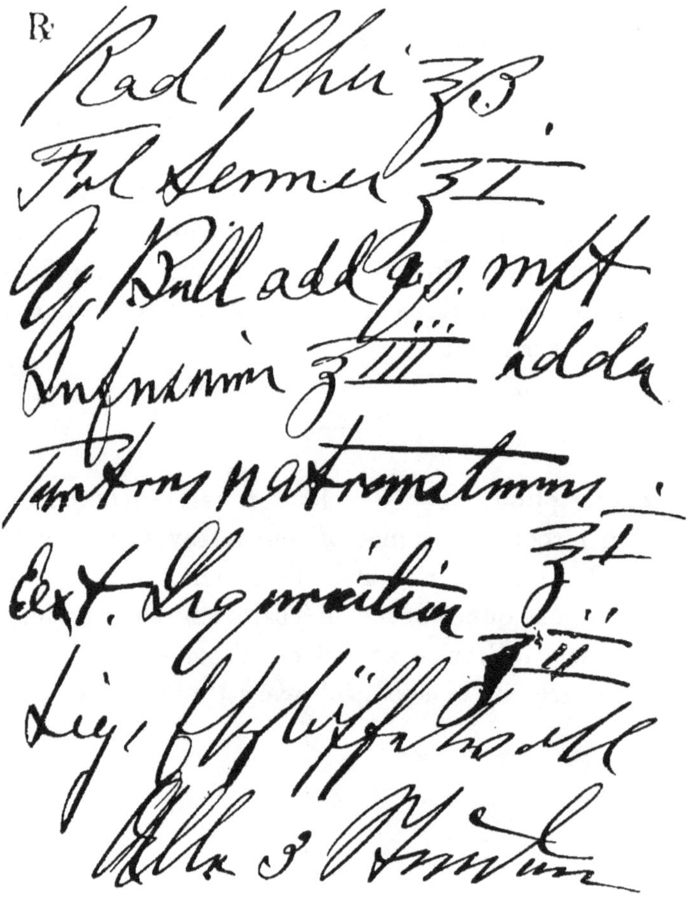

DECOCTIONS.

Decoctions differ from infusions only in the fact that the drug is boiled in the water. They should be made the same strength as infusions when not otherwise directed, boiled fifteen minutes and strained.

COMPOUNDING EMULSIONS.

Emulsions are aqueous liquid preparations in which oily or resinous substances are suspended by the use of gummy or viscid substances. Gum arabic is the best emulsifying agent. These give the young pharmacist more trouble than any other class of prescription. Physicians often order insufficient gum or otherwise render it impossible to secure a good emulsion by following their directions. In such cases the dispenser should use his own judgment and follow his own methods in compounding the prescription as long as the physiological effects of the medicine are not altered. When alcoholic ingredients are to be added to the emulsion let them be diluted as much as possible and added after the emulsion is nearly finished. On the other hand alkalies or alkaline solutions assist in emulsifying the oil and can be added while forming the emulsion. There are two distinct methods of making emulsions, either of which give good results if followed closely. The first method is to take the ingredients in the following proportions: Gum one, water two, and oil four. Place the gum in a dry mortar, add the oil and stir with the pestle until they are thoroughly mixed, then add the water and upon stirring again the emulsion is very quickly made. After this nucleus is formed the rest of the water and other ingredients may be added. The other method is to place the gum in a mortar and add gradually sufficient water to make a thick smooth mucilage. To this the oil is added with constant stirring and from time to time a few drops of water may be added if the mixture becomes too thick. If the finished emulsion is not perfect or shows signs of oil globules it may be remedied by placing more gum in a dry mortar and gradually adding the emulsion with constant stirring. Gum-resin emulsions are made by simply rubbing the gum resin in a mortar with a small quantity of water until a smooth uniform paste is formed, then adding the remainder of the water and straining. Seed emulsions are made in the same manner, the emulsifying agent being the gummy substance contained in the seed itself. Emulsions are sometimes made with condensed milk. It is excellently adapted for making emulsions of any kind. A 50 per cent. cod liver oil emulsion is thus made with it: Oil, 8.0; condensed milk, 3.0; glycerin or syrup, 3.0; water, 2.0. The milk is rubbed in a mortar, the oil added gradually and lastly the glycerin and water.

The following prescription is often met with:

Ol. ricini.................two drachms
Sacchari.................. "   "
Mucil-Acaciae............. "   "
Aquæ ....one ounce
Ol. menth. pip. gtt. ij.

This is best mixed in a mortar; rub the oil of peppermint with the sugar, add the mucilage and a little water, then the oil, and when these are well mixed, the remainder of the water gradually; you will then have a nice milky-looking mixture, without any globules of oil floating about. Always take care that the mucilage and oil are well mixed in this kind of mixture before the water is added, or you will have drops of oil floating about, and should any tincture or spirit form part of the ingredients, mix it with a little of the water, and let it be added last, or you may possibly find the mixture "come unmixed," for gum is precipitated from its solution by spirit; and do not forget that the oil is to be added to the mucilage, not the mucilage to the oil.

FIG. 14.

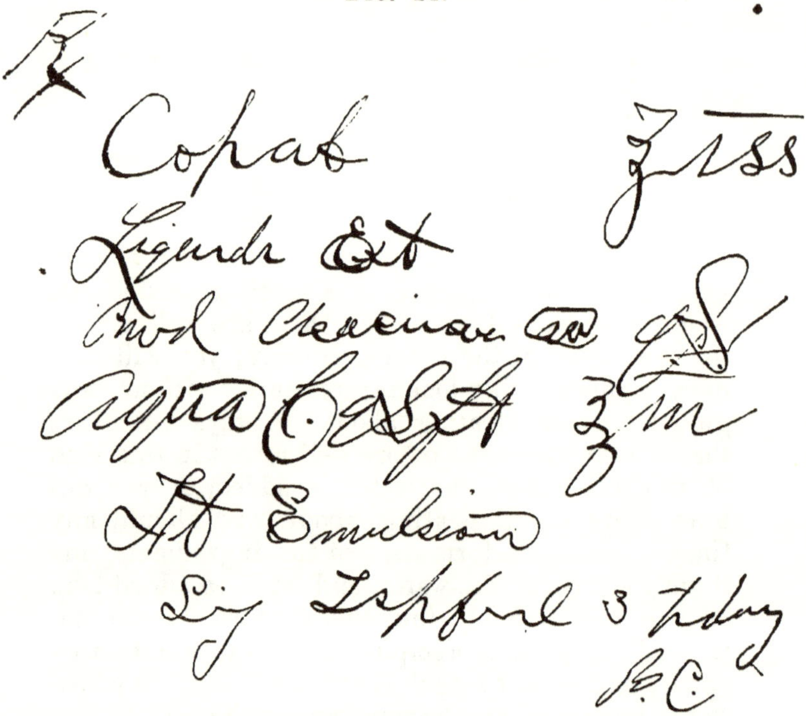

Prescription No. 14 should be prepared as follows:

Place in a dry mortar six drachms of powdered gum arabic and as much of extract of licorice, thoroughly dry, and pour on the balsam of copaiba. Mix well, and add at one time twelve drachms of camphor water. Continue the stirring with the pestle till the mixture is thoroughly homogeneous, scraping now and then the side of the mortar and

—44—

the pestle, so that no balsam can escape emulsion.
Now add more camphor water by small portions at a
time, and finally complete the three fluid ounces as
prescribed.

Emulsions of turpentine and most volatile oils
can be made as follows: One ounce of oil is placed
in an eight ounce bottle, one half ounce of powdered
acacia is added, and the bottle shaken well, until
they are thoroughly mixed, then add two ounces of
syrup and sufficient water to make eight ounces.
The emulsion is completed by a thorough shaking.

PILLS.

Pills are small, solid bodies, of a globular, ovoid,
or lenticular shape. Pharmacists, however, always
make them globular when compounding prescriptions specifying pills to be made. The ingredients
are first reduced to a soft or pulverent condition and
then made into a pill mass of such consistency that
it may be easily rolled into a cylindrical form for the
purpose of dividing it into the required number of
parts. The first step is to thoroughly mix the ingredients. To do this all dry substances should be
reduced to a very fine powder and triturated well together. Some exceptions to this rule are where a
very minute quantity of a powerful ingredient enters
into the prescription it is sometimes advisable to dissolve it in a few drops of solvent in order that it may
be more evenly divided throughout the mass. Phosphorus and strychnine are such drugs. Again,

solid extracts are often too tenacious to powder, and require softening by a solvent before they can be intimately mixed with the other ingredients. After the ingredients are thoroughly mixed the next step is the selection of the excipient with which to form them into a mass. This is a very important point and requires a thorough knowledge of the physical properties of all of the ingredients. Water can often be used where the ingredients have sufficient adhesiveness in its presence.

Syrup of acacia is used where considerable adhesiveness is required.

Glucose is still more adhesive.

Soap makes a good excipient for resinous substances.

Cocao butter and resin cerate form valuable excipients for oxidizable substances.

The following is a good general excipient for pills. It should be kept on hand in small quantity and in convenient form for use:

Pulv. Acacia.................45 grains
Glycerine.....................½ oz. av.
Glucose......................2 oz. av.

Add the acacia to the glycerine, mix well, then add the glucose and apply a moderate heat until the acacia is dissolved.

In forming the mass it should be made adhesive, firm and plastic. We mention below some excipients best adapted to special cases.

When aloes in any form enters as an ingredient in a formula for pills, an excellent mass may be worked up on the addition of a few drops of decoct. aloes co. Most resinous extracts and gum resins are formed into a good mass with a little mucilage or spirit. For ipecacuanha, rhubarb and powders of this class, syrup forms a good excipient. Powdered rhubarb may also be made into a good mass with thin honey. Tincture of jalap may be used as an excipient of powdered jalap. Euonymin, leptandrin, iridin, and drugs of this class mass well with glycerite of tragacanth. For hard extracts heat is often of great assistance, and a warm mortar or pill tile will be found very useful in bringing them to a plastic consistence. When it is necessary to make very soft extracts, such as cascara sagrada or viburnum prunifolia, alone into pills, and the addition of powdered gum would render them too bulky, the extract should be evaporated down over a water bath until almost dry. Should the extracts be already hard and dry they may be reduced to powder and worked into a mass with a few drops of spirit. When essential oils are prescribed alone, a good mass may be formed with calcined magnesia and a small quantity of soap. Most intractable ingredients may be worked into a fair mass by the aid of glycerite of tragacanth, or the mixture of tragacanth and treacle with a little powdered gum if necessary.

Antipyrine is made into a good pill with glycerite of tragacanth, or with powdered gum and water.

Argentic nitrate with kaolin ointment, sugar of milk or manna.

Benzoic acid with Canada balsam, 1 to every 4 grains, or with glycerine, 1 drop to 5 grains.

Balsam of Peru with bread crumb or beeswax.

Calomel with confection of roses, or manna and compound tragacanth powder. Calcined magnesia should not be used with calomel.

Camphor.—The gum having first been reduced to a very fine powder, it may be worked up with glycerite of tragacanth and soap, castor oil and soap.

Camphor monobromata, with Canada balsam, 1 grain to 5, in a warm mortar.

Carbolic acid, with powdered licorice, 1 grain, to each minim and mucilage. A firm pill may be formed with powdered altheæ and glycerine in the following proportions: Acid carbolic, 2; pulv. altheæ, 3; glycerin, ¼. Another method is with powdered soap, 1; powdered licorice, 5; acid carbolic, 1. This when properly worked makes an excellent mass.

Cerium oxalate with glycerite of tragacanth or confection of roses.

Chloral hydrate with Canada balsam, ½ grain to 5, or syrup and powdered tragacanth.

Creosote.—Several methods are employed for making a pill mass, the success in forming the pill depending a good deal on the manipulation.

1. With calcium phosphate and hard soap.

2. With powdered licorice and glycerite of tragacanth.

3. With bread crumb, 2 to 1.

4. With powdered soap, 1 part; licorice in powder, 5 parts; creosote, 1 part.

Copaiba balsam.—When mixed with calcined magnesia and allowed to stand for a length of time a workable mass is formed. Carbonate of magnesia or slack lime answers very well. Also with calcined magnesia and beeswax.

Essential oils, such as savin, cloves, etc., may be massed with calcined magnesia and powerded soap, or with calcium phosphate and soap. Soap and powdered licorice also make a good base, 1 to 5.

Croton oil with bread crumb, magnesia and soap, or powdered licorice and mucilage.

Extract canabis indica and other thin extracts may be massed with compound tragacanth powder and magnesia.

Hydrarg. c. creta with confection of roses. Care should be taken not to work it too hard, or the mercury will separate from the chalk.

Pepsin with glycerine and powdered tragacanth.

Potassium iodide, bromide and other crystalline salts should be reduced to a very fine powder, and massed with a small quantity of licorice powder, powdered tragacanth and a drop of water.

Potassium permanganate, with kaolin ointment or resin ointment, etc., decomposes when mixed with organic substances.

Quinine with glycerite tragacanth, 1 to 4, also with one drop of dilute sulphuric acid to every 5 grains.

A good pill is made with tartaric acid, 1 grain to every 10 grains of quinine, and a drop of water. Also with lactic acid, 3 minims to every 16 grains, and confection of roses and glycerine.

Tannic acid with glycerine, and a little powdered tragacanth if necessary.

Thymol should be reduced to a fine powder, mixed with powdered soap, and massed with a drop of rectified spirit.

FIG. 15.

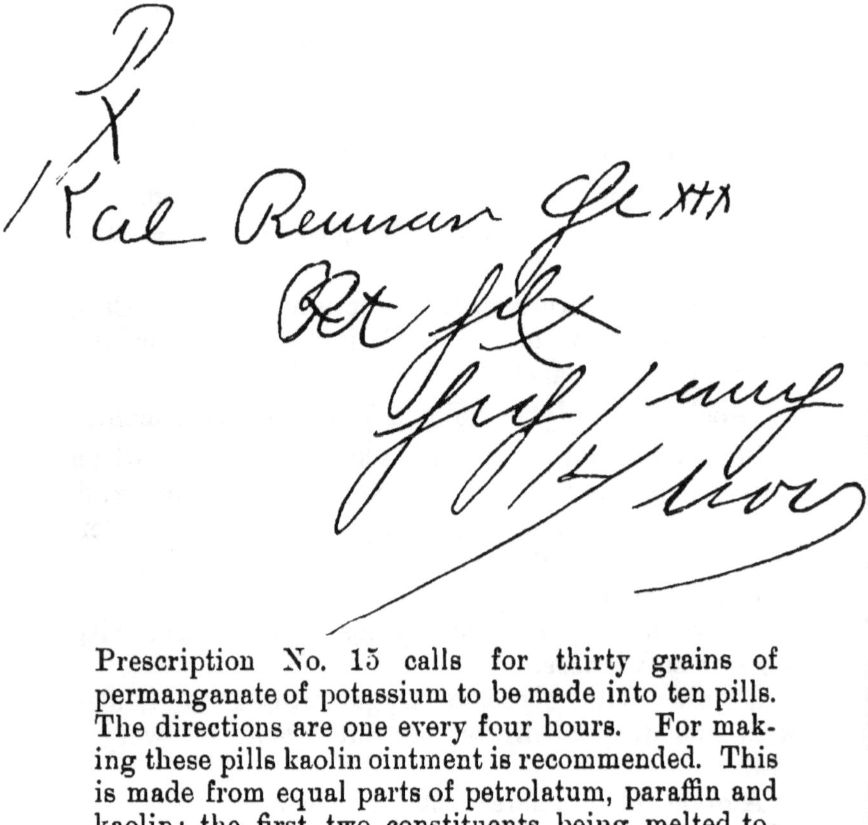

Prescription No. 15 calls for thirty grains of permanganate of potassium to be made into ten pills. The directions are one every four hours. For making these pills kaolin ointment is recommended. This is made from equal parts of petrolatum, paraffin and kaolin; the first two constituents being melted together, then kaolin added and stirred in until cool. It is said to be the only pill-mass which has ever been successfully used with potassium permanganate.

Pill masses should be made in a mortar specially adapted to that use. Figure 16 is a mortar made in  the most approved shape for working pill masses. The pestles are made with the wedgewood part very much longer than in the ordinaiy pestles, so that when in use the

wood of the handle will not rub on the edge of the mortar. They are not only cleaner themselves than the old style pestle, but the mass in the mortar is kept cleaner.

After the mass is properly prepared it is placed upon a pill tile or pill machine and rolled into a long cylinder of the proper length and then cut into the required number of pieces with a spatula or the cutter attached to the pill machine. Figures 17 and 18 show two styles of pill machines made by Jno. M. Maris & Co., of Philadelphia.

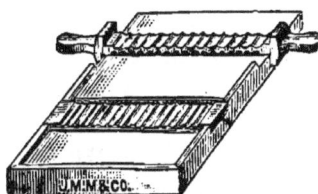
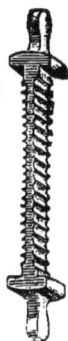
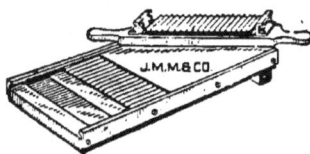

FIG. 17.

FIG. 18.

While rolling the mass upon the tile or pill machine they should be dusted freely with rice flour. powdered magnesia or lycopodium to prevent sticking. After the pills have been cut they can be par' tially shaped by rolling backward and forward between the cutting edges of the machine; afterward they may be finished by rolling between the fingers or by rolling them upon a level surface under the pill finisher shown in figure 19. When the pills are dispensed uncoated a small quantity of dusting powder should be placed in the box to prevent adhesion of the pills.

FIF. 19.

Pill Coating.—Some physicians and pharmacists prefer to have all the pills prescribed or dispensed by them coated with the view of masking their taste or giving them a finer appearance. The dispensing pharmacist generally uses gold or silver leaf, or gelatin for this purpose.

Gelatin Coating.—The following makes a good solution:

Gelatin........................... 1 oz.
Water........................... 7 ozs.

Dissolve at a gentle heat, then add the white of an egg, and heat until the albumen coagulates, strain through a flannel into a water-bath kept at a low temperature, add 2 drams glycerin, 2 drams alcohol and 6 grains of boric acid.

A large round cork can be brought into use by setting upright in it 6 or 12 needles and impaling upon the point of each one of the pills which are now dipped into the warm solution, taking care not to keep them in too long, as a thick coating is not desired. After removing them from the solution they should be turned about in different directions to render the coating even and then set aside to dry.

Gold or Silver Coating.—This is a very simple process and is accomplished by placing a drop or two of syrup of acacia in a mortar and after distributing it well over the surface, the pills are put in and rotated so as to coat each one with a thin layer of the mucilage, after which they are put into the coater with the gold or silver leaf and rotated gently until perfectly coated.

Fig. 20 shows a good and cheap style of pill coater.

Pills are seldom sugar coated by the dispensing pharmacist as the process requires considerable time when done on a small scale, and the gelatin coating answers all practical purposes.

FIG. 20.

FIG. 21.

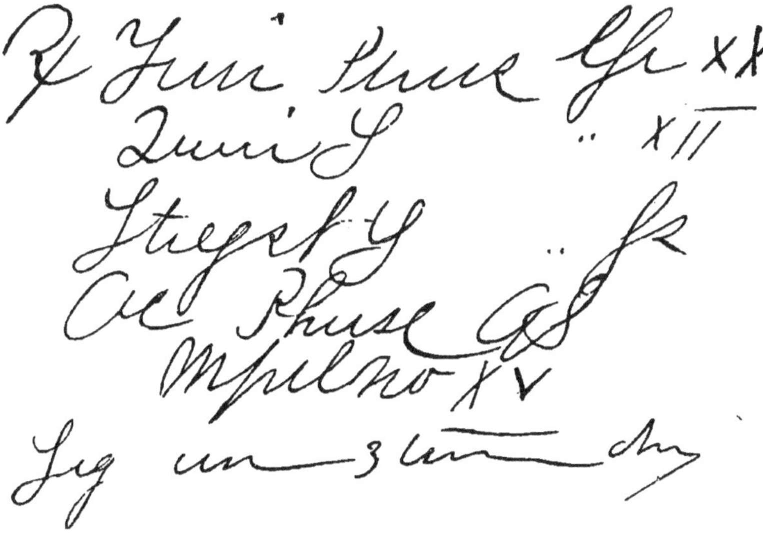

Prescription No. 21 would probably give some trouble in reading it to those unacquainted with the doctor's style. It is translated as follows:

Ferri. Phosph............Grains XX.
Quinine Sul..............Grains XII.
Strych. Sul...............Grain ss.
Acid Phos. Conc...............Q. S.
M. pil.....................No. XV.
Sig.—One 3 times a day.

The strychnine should be finely powdered and carefully triturated with the phosphate of iron until they are thoroughly mixed. The quinine may now be added and the whole again well triturated. The mass is made with syrupy phosphoric acid, and as it acts as a powerful solvent it must be used with caution, about 18 or 20 drops being usually sufficient. The ingredients should be massed rapidly, rolled and divided into pills without delay, or they will soon become too hard to mold.

SUPPOSITORIES.

Suppositories are solid conical bodies with a rounded apex, intended to be introduced into the rectum, vagina or urethra, and of such consistency as to retain their shape at ordinary temperatures but soften or melt at the temperature of the body. Cocao butter and a mixture composed of glycerine and gelatin are the two bases more commonly used in making suppositories. When the weather is extremely warm or when such ingredients as volatile oils, chloral hydrate, or carbolic acid enter into the suppositories, it is generally necessary to raise the melting point of the cocao butter by the addition of a little spermaceti or white wax. This should be avoided whenever possible. When the size of the suppositories are not directed in the prescription they should be made to weigh fifteen grains. Suppositories are shaped in three ways, viz—rolling, moulding or pressing. The first method has the advantage of requiring no special apparatus, and with a little practice the pharmacist can soon turn out very fair suppositories. The ingredients are first reduced to fine powder, or if there be an extract it should be rubbed in the mortar to a smooth paste with a few drops of water, the cocao butter is then added and the whole made into a mass, which is rolled into a cylinder on a pill tile and cut into the required number of pieces. The tile should be dusted freely with lycopodium or powdered elm bark while rolling the suppositories upon it. Each piece is then partially shaped with the fingers and finally finished by rolling with a spatula. If the mass is too hard or brittle to roll well the heat of the hand is generally sufficient to soften it, if not a drop or two of olive oil will suffice. For dispensing suppositories the pharmacist should be supplied with boxes having separate compartments for each one, which prevents injury to the suppositories. Such boxes may be secured from any wholesale druggist. If an ordinary box is used the suppositories should be dusted with lycopodium and protected by a layer of cotton.

Compressed suppositories are made by passing the mass prepared as for rolling, through a machine which shapes them by pressure. These machines are generally expensive and very few of them satisfactory.

Gelatin suppository capsules are sometimes used. The ingredients of the suppository are placed in the lower portion and the cap which is conical is adjusted after dampening the upper end of the lower portion with water. Suppositories should never be dispensed in capsules unless especially ordered.

Moulded suppositories are made by mixing the medicinal ingredients thoroughly with a small quantity of the cocao butter by rubbing in a mortar and adding this mixture to the remainder of the cocao butter previously melted in a casserole and cooled to about 95 degrees F. Mix thoroughly by stirring and pour into the moulds, which should be kept cold by being placed upon ice. Where extracts are present in the mixture the least possible amount of heat should be used. The best plan is to heat the casserole by holding it in a pan of hot water, thus avoiding direct heat which evaporates the water and leaves the extract to coagulate and separate from the mixture. Figure 22 is a casserole of good size and shape for

FIG. 22.

suppositories. There are many styles of suppository moulds. Figure 23 represents the individual mould. It is often difficult to remove the suppository from this style of mould. Figure 24 is better in that respect as it opens, exposing one-half of each suppository.

Fig. 23.

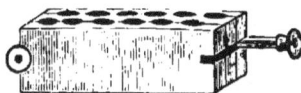

Fig. 24.

Fig. 25 we consider the best and most practical suppository mould now on the market. It is well and substantially made of brass, nickle-plated. Fig A represents the mould when closed, and Fig. B when opened.

On the bottom of the inner ring there is a projection of about $\frac{1}{8}$ of an inch. This is intended to be pushed down on the counter after the set-screw is loose, by bearing down on the outer ring, thus starting the inner ring, so it can then be easily pushed up with the hand. This mould is made in different sizes.

Fig. 25.

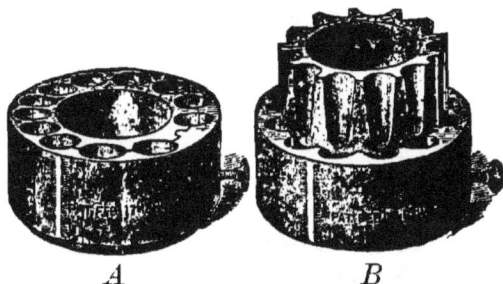

A   B

Gelatin suppositories are made by using one of the following mixtures as a base :

Glycerine-Gelatin.—Hard.

Gelatin...........................25
Water.............................70
Glycerine.........................50

Macerate the Gelatin in the water for several hours; add the glycerine, then heat on a steam bath until the mass weighs 100.

Glycerine-Gelatin.--Soft.

Gelatin.............................15
Water.............................45
Glycerine..........................50

Proceed as directed in the forgoing formula.

The base used must be chosen according to the nature of the ingredients, if they are liable to render it thin or unctious the hard base should be used and if necessary a little powdered tragacanth can be added, if the ingredients should have an opposite effect, the soft base should be used. The base is melted by means of a water bath and the medicament added in concentrated solution, or in fine powder (especially if it be soluble in hot glycerine) and then poured into the moulds.

FIG. 26.

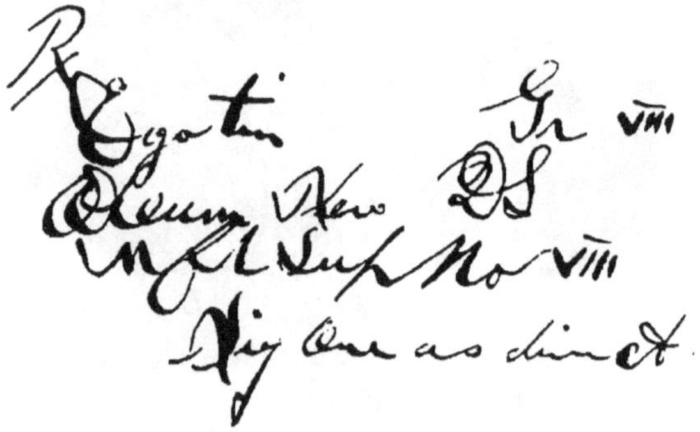

Prescription No. 26 calls for 8 grains of ergotine and a sufficient quantity of cocao butter to make 8

suppositories. The ergotine should be rubbed to a smooth thin paste with a drop or two of water, then mixed with about $\frac{1}{3}$ of the cocao butter, after which it is added to the rest of the cocao butter which has been melted in a casserole, and stir until all are well mixed. The least possible amount of heat should be used.

CAPSULES.

Of late the use of gelatin capsules for administering nauseous or bitter medicines has become quite general. There are two methods of filling them in vogue. The first is to place the ingredients in the capsule, in the form of a powder. This is done by mixing the ingredients and reducing them to a powder, which is divided with a spatula into the proper number of parts and placed in the capsules, either by the use of a capsule-filler or by pressing the longer end of the empty capsule into the powder until it is picked up, or using a spatula to force the powder into it. The other method is to make a mass of the ingredients, divide into the proper number of parts and make little rolls of them, which are then placed in the capsule.

FIG. 27.

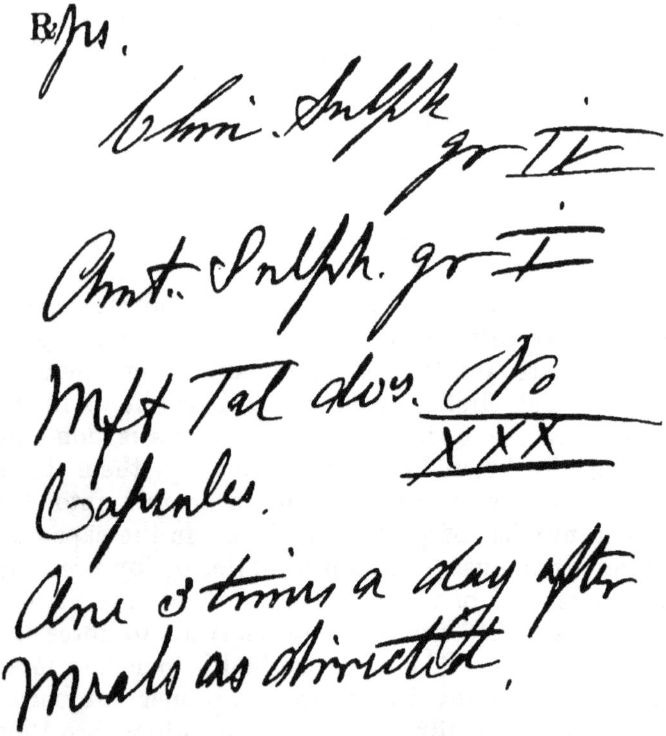

Prescription No. 27 is very well written, and
would give no trouble provided one is acquainted
with the two Latin words, "tales doses," meaning of
such doses. In this prescription the quantities of
the ingredients for one dose are given and the pharmacist instructed to make thirty such doses and put
them in capsules.

Prescription No. 28 is written in the ordinary hand
of a physician, who often does much better than this.
One drachm of salicylate of cinchonidia is ordered
to be made into twenty capsules. The directions
are, one four times daily.

POWDERS.

When medicines are ordered to be dispensed in the form of powders, the ingredients should be finely powdered and intimately mixed. The papers for wrapping them in should all be creased at once, by folding down a narrow margin at one side, they should then be arranged in order in front of the operator with the crease from him. The mixture is then taken upon a piece of paper, flattened out and arranged into a square which can be divided into the required number of parts with a spatula. They are then transferred to the papers and folded, care should be taken to have each powder uniform in size and shape. After the powder is folded the ends can be bent over using the box which is to hold the powders as a guage. FIG. 29. However, Fig. 29 is a much better guage for this purpose and it can be changed to suit any size powders.

FIG. 30.

Prescription No. 30 calls for thirty grains of powdered camphor; (impalpable) reduced iron, ten grains; Misce. Ft. Chart., 20. Sig.—One three times a day. The camphor should be powdered very fine, a few drops of alcohol being added which will soon evaporate, next add the reduced iron, mix well and divide.

5

FILING AND PRESERVING PRE-
SCRIPTIONS.

After a prescription has been compounded it should be dated, numbered and placed upon a file for preservation for future reference, or refilling if required. A corresponding number and date having been placed on the label of the medicine serves to identify the prescription when it is again needed.

Figure 31 represents the Dufalt Safety Prescription File which is the most convenient one for daily use. It can be bought of any wholesale drug house. Such a file should be kept upon the prescription case for receiving the new prescriptions. There are many methods of filing and preserving the prescriptions after being taken from this temporary file, one of which is to paste them in a book, another is to copy the prescriptions into a book, but this plan cannot be recommended, as copying in-

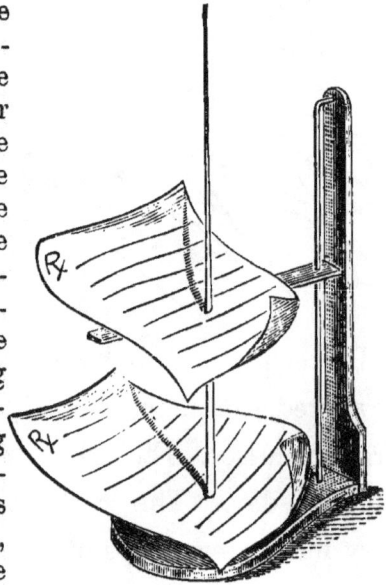

FIG. 31.

creases the chances of errors and the originals would still have to be kept as proof of the correctness of the copy. The better plan is to keep each 500 prescriptions on a file provided with a box-covering to protect them from the dust and dirt; on the outside of the cover should be marked the first and last number of the prescriptions contained in it, also the dates.

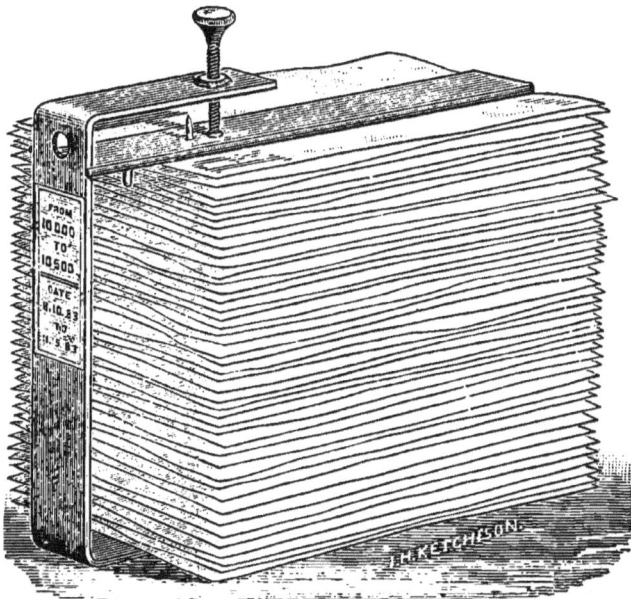

FIG. 32.

Figure 32 shows the Nesbitt prescription file which is a very handy and convenient one for preserving prescriptions.

THE PRESCRIPTION CASE.

We will leave the general arrangement of the prescription case to the individual ideas of each pharmacist, as there is no uniformity of opinion on the matter, and mention a few points which contribute greatly to convenience and accuracy in compounding. A separate closet should be attached to the case for very powerful and poisonous drugs, and the pharmacist should avoid keeping two drugs with similar names in close proximity to each other. The most important feature of the prescription case is the prescription scale.

All prescription departments should be furnished with at least two pairs of prescription scales, one to be kept for weighing small quantities (never over twenty grains), and the other for larger quantities up to the ounce.

By keeping a scale for small quantities, its delicacy will be retained for a very much longer time than if used for all weights, heavy and light.

In too many pharmacies can be seen prescription scales that will not turn for a quarter of a grain; this is due to dull bearing points, too large a weight having been used on some occasion, or to rust or dirt being allowed to collect on them.

The prescription scale should be cleaned with water, and if care is exercised nothing else is needed.

They should always be enclosed in a case, protected from the air and dust. Always see that your scales balance before attempting to weigh.

It is well to place pieces of paper of even weight on the pans, for by this means you avoid the danger of soiling the latter, and the substance weighed can be at once carried to where you wish to deposit it.

Keep the case door closed when the scale is not in use; put away the weights after using them, and when handling weights, use a little pair of nippers.

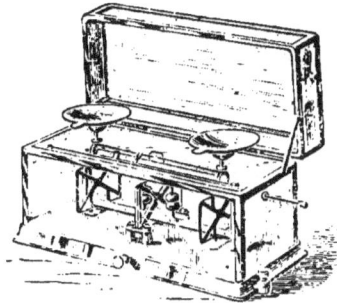

FIG. 33.

Fig 33 shows the "Favorite" torsion balance prescription scale. Its capacity is 8 ounces but it is sensitive to 1-64th grain. This is undoubtedly the best prescripscale on the market.

All graduates used at the prescription case should be tested before used, as many found on the market vary from their markings.

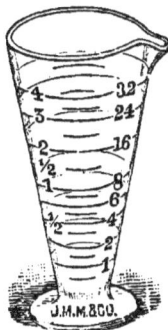

FIG. 34.

FIG. 35.

Figs. 34 and 35 are graduates made by Jno. M. Maris & Co., and will always be found accurate. Every pharmacist should have this minim graduate, instead of counting drops as minims as so many do. Convenient to the compounder should be one or more general excipients for pill making, kept in convenient form for use, also several bottles of dusting powder such as

—67—

rice-flower, lycopodium and powdered licorice. The most convenient form for keeping these is in the common glass individual pepper boxes.

A small alcohol lamp is another great convenience. Fig. 36 shows a good one. Fig. 37 is a good style of cork press which every pharmacist should have.

Conveniently at hand should be kept all the latest books of reference, and at least one or two good journals of pharmacy, such as The

FIG. 36.

FIG. 37.

Pharmaceutical Era or The Druggist Circular. No pharmacist can keep up with the times without such literature. After being read these should be carefully preserved for binding and will become very valuable for reference and repay many times their cost.

EXPLOSIVE PRESCRIPTIONS.

The following prescriptions contain most of the substances which have been found to produce explosions. Potassium Chlorate, and in fact all other Chlorates, should never be dispensed with organic, combustible or oxidizable bodies.

A mixture of Hypophosphite of lime, Chlorate of Potassium, and Lactate of Iron exploded, and nearly killed the prescription clerk who was compounding it.

Even the simple trituration of Calcium Hypophosphite is dangerous. A young pharmacist was killed by an explosion which was caused by the shakingof a solution of this substance. Physicians not infrequently order a solution of Chromic acid in glycerine. But when the acid is added quickly and all at once to the glycerine, a readily explosive substance like nitro-glycerine is formed. Chlorate of potassium, when mixed with tannin or muriate of morphia, often explodes. The combination of iodine and preparations of ammonia must be made cautiously, as iodide of nitrogen is formed, which explodes on the slightest touch. Indeed, one ought to be very careful in ordering and compounding mixtures in which easily reducible substances enter, such as the chlorates, the hypophosphites, the nitrates, preparations of iodine or ammonia, chromic acid, glycerine, permanganate of potash, etc.

The following, taken from physicians' prescriptions, are dangerous, and have caused serious accidents.

R. Potassa Chlorate.
   Pulvis Cateohu.
M.

R. Potassa Chlorate,
   Sodii (or Calcii) Hypophosphite, Aqua.
M. Dissolve the two salts separately or an explosion will occur.

R. Potassa Permanganate,
   Glycerine.
M. This is almost sure to cause an explosion.

R. Acid Nitric,
   Acid Muriatic,
   Tr. Nux Vomica.
M. Exploded in about two hours.

The following prescription can not be prepared without an explosion:

R. Lactis Sulph.
   Antimonii Sulph. Aurant, aa..... Gr III
   Zinci Valeri .... " I
   Patass. Chlorate. .... " II
M. Ft. Pulv. Deutur. doses, tales, No. XII

R. Argenti Oxide,
   Morphia Muriate,
   Ext. Gentian.
M. This mixture has exploded.

R. Turpentine,
Acid Sulphuric.

M. Mix slowly in a large, open vessel.

R. Potassa Permanganate,
Alcohol,
Aqua.

M. Mix the alcohol and water; add the potash
slowly and cork loosely.

R. Potassa Chlorate,
Acid Tannic,
Glycerine,
Aqua.

M. Dissolve the Tannin in the Glycerine, the
Potash in water, and mix.

R. Potassa Chlorate,
Tr. Ferri Chloridi,
Glycerine.

M. This is liable to cause an explosion when
warmed.

R. Soda chlor.,
Antim. Sulph. Aurat.

M. Takes fire even when triturated very gently.

ABBREVIATIONS.

```
A, aa............... Anna............Of each.
Ad............................. To, up to.
Ad duas vices.................. At twice taking.
Ad secundum vicem........... To the second time.
Adde or addantur................. By adding.
Ad gratam aciditatem.... To an agreeable sourness.
Ad libitum....................... At pleasure.
Adstante febre............. When the fever is on.
Agitato vase................... Shake the bottle.
Alter ............................ The other.
Alternis horis................. Every other hour.
Amplus ............................ Large.
Aqua Bulliens................... Boiling water.
   "    Communis................ Common water.
   "    Fervens................. Hot water.
   "    Fontalis............... Spring water.
   "    Pluvialis.............. Rain water.
Aut ................................. Or.
Balneum arenæ.................. Sand-bath.
   "    Vaporis................. A vapor bath.
Bene............................... Well.
Bibe ............................. Drink.
Bis ............................. Twice.
Bis in die....................... Twice a day.
Bolus.......................... A large pill.
Calefactus ...................... Warmed.
Caute .......................... Cautiously.
```

Charta...Paper.
Cochleare...............................A spoonful.
    "   Amplum.................A tablespoonful.
    "   Magnum................A large spoonful.
    "   Medium...............A desert spoonful.
    "   Parvum.................A teaspoonful.
Cola.......................................Strain.
Coletur.............................Let it be strained.
Collyrium.........................An eye-wash.
Concisus.......................................Cut.
Congius....................................A gallon.
Continuantur remedia...... Continue the medicine.
Contusus..................................Bruised.
Cortex....................................The bark.
Cum...With.
De.......................................Of or from.
Decanta....................................Pour off.
Decem..Ten.
Decubitus..............................Lying down.
Dein...................................There upon.
Diebus Alternis..................Every other day.
Dimidius..................................One half.
Dividendus........................To be divided.
Dolor...Pain.
Dosis...A dose.
Ejusdem...............................Of the same.
Emesis....................................Vomiting.
Enema.....................................A clyster.
Et...And
Fiat, Fiant....Let it be made, Let them be made.
Farina ....Flour.
Febre durante..................During the fever.
Fiat cataplasma.................Make a poultice.

Fiat emplastrum	Make a plaster.
" " vesicatorium	Make a blister.
" enema	Make an injection.
" haustus	Make a draught.
" massa	" " mass.
" secundum artis	Make according to art.
Gargarisma	A gargle.
Gradatim	Gradually.
Gutta	A drop.
Guttatim	By drops.
Hora	An hour.
Hora somni	At bed-time.
" decubitus	At bed-time.
Horis intermediis	In the intermediate hours.
Infunde	Pour in.
Inter	Between.
Magnus	Large.
Manipulis	A handful.
Mica panis	Crumb of bread.
Misce	Mix.
Mittatur	Let it be sent.
Non	Not.
Numerus	Number.
Octarius	A pint.
Omni hora	Every hour.
" mane	Every morning.
" nocte	Every night.
Partitis vicibus	In divided doses.
Parvus	Little.
Per	Through, by.
Pone aurem	Behind the ear.
Potus	Drink.
Pro	For.

Pro re nata........................Occasionally.
Pugillus................................A pinch.
Quantum libet............As much as you please,
    " sufficiat............A sufficient quantity.
Semis........................ ....A half.
Sine.... .... Without.
Talis..........................Such, like this.
Ter.............................Three times.
Ter in die....................Three times a day.
Tere..Rub.
Vel.................................. ....Or.

SOLVENTS.

Acid Boric . . . . Glycerine and hot water.
  "   Arsenious . . . . Boiling water, hydrochloric acid.
  "   Tannic . . . . Glycerine.
Alkaloids . . . . Dilute Acids.
Antipyrin . . . . Water, alcohol.
Antifebrine . . . . Water.
Camphor . . . . Alcohol.
Corrosive Sublimate . . . . Alcohol.
Chloralamid . . . . Alcohol.
Gums . . .   . Water.
Gum Resins . . . . Alcohol.
Gum Cotton . . . . Acetic Ether.
Gutta Percha . . . . Chloroform.
Iodine . . . . Alcohol.
Iodoform . . . . Oil of cassia.
Iodol . . . . Water.
Phenacetine . . . . Alcohol.
Phosphorus . . . . Ether and fatty oils.
Quinine . . . . Dilute acids.
Resins . . . . Alcohol.
Strychnine . . . . Dilute acids.
Sulphur . . . . Bisulphide of Carbon.
Salol . . . . Alcohol.

TOXICOLOGY.

There are three classes of poisons, viz: Irritants, which produce irritation and inflammation in the stomach.

Narcotics, which effect the brain and spinal cord, producing headache, giddiness and insensibility.

Narcotic Irritants, having the double action.

ANTIDOTES.

In cases of poisoning by alkalies give vinegar, oil and milk freely and produce vomiting.

For acids, give chalk, soda, lime-water and demulcent drinks of flax-seed or slippery elm, then a prompt emetic.

In all ordinary cases of poisoning (unless by strong acids or alkalies), first give an emetic, fifteen grains of Sulphate of Zinc. If not at hand, give a tablespoonful of ground mustard in a glass of water. Repeat this every few minutes until the stomach is emptied. Then follow with white of eggs, milk, or chemical antidote as given below.

Acid, Carbolic—The best antidote is a mixture of olive oil and castor oil, freely administered, or a mixture of slacked lime with about three times its weight of sugar rubbed together with a little water.

Acid, Prussic—When taken in large doses is generally immediately fatal. In smaller doses the symptoms are weight and pain in the head, giddiness, nausea, rapid pulse, loss of muscular power, foaming at the mouth.

Antidote—Cold water to the head and spine, Carbonate of Ammonia internally, small doses of Chloride of Lime or Soda. The chemical antidote is moist Peroxide of Iron.

Acid, Sulphuric—Carbonate of Soda, Lime and Magnesia. Water must not be given.

Aconite—Whisky and prompt Emetics.

Alkaloids—In cases of poisoning by alkaloids, emetics and the stomach-pump must be relied on rather than chemical agents. But astringent liquids may be administered, for tannic acid precipitates many of the alkaloids from their aqueous solution, absorption of the poison being thus possibly retarded.

Antimony—The introduction of poisonous doses of antimonials into the stomach is fortunately quickly followed by vomiting. If vomiting has not occurred, or apparently to an insufficient extent, any form of tannic acid may be administered (infusion of tea, nutgalls, cinchona, oakbark, or other astringent solutions or tinctures) an insoluble tannate of antimony being formed, and absorption of the poison consequently somewhat retarded. The stomach-pump must be applied as quickly as possible.

Arsenic—Hydrated Oxide of Iron.

Atropia—Coffee, Tannin, Opium.

Belladona—Coffee, Tea, Jaborandi, Opium.

Blue Stone—Milk and white of eggs in large quantities.

Chloral Hydrate—See Chloroform.

Chloroform—Horizontal position, cold water to the head, stimulants.

Digitalis—Coffee, Brandy, Ammonia.

Gelsemium—Coffee, Brandy, emetics of Ipecac.

Opium—Coffee, Atropia hypodermically, emetics; keep the patient in motion.

Silver—Solution of common salt, sal-ammoniac, or any other inert chloride should obviously be administered where large doses of nitrate of silver have been swallowed. A quantity of sea-water or brine would convert the silver into insoluble chloride, and at the same time produce vomiting.

Strychnine—Chloral Hydrate, 10 grains every 15 minutes; Ether and Opium.

Sugar of Lead—Epson Salts, Lemonade.

Salts of Mercury—Milk, white of eggs, emetics.

The arsenical antidote (Hydrated oxide of iron) is prepared as follows:

To Liquor Ferri Tersulphatis add excess of Aqua Ammonia, collect the precipitate on a filter, and wash well with water.

WEIGHTS AND MEASURES.

TROY OR APOTHECARIES' WEIGHTS.

20 grains = one scruple = 20 grains
3 scruples = one drachm = 60 "
8 drachms = one ounce = 480 "
12 ounces = one pound = 5760 "

AVOIRDUPOIS WEIGHTS.

437.5 grains = one ounce = 437.5 grains
16 ounces = one pound = 7000 "

APOTHECARIES' OR WINE MEASURE.

60 minims = one fluidrachm
8 fluidrachms = one fluid ounce
16 fluid ounces = one pint
8 pints = one gallon.

IMPERIAL MEASURE.

60 minims = one fluidrachm
8 fluidrachms = one fluid ounce
20 fluid ounces = one pint
8 pints = one gallon

DOMESTIC MEASURES.

A teaspoonful is equal to 1 fl drachm
A desertspoonful " " " 2 " drachms
A tablespoonful " " " 4 " "
A wine glass " " " 2 " ounces
A teacup " " " 4 " "

METRIC WEIGHTS.

The units of the Metric System of weights and measures are the Metre, Litre and Gramme.

The metre which is the unit of length is the $\frac{1}{40000000}$ part of the earth's circumference around the poles and its length is 39.370432 inches.

The litre or unit of capacity is the cube of $\frac{1}{10}$ part of a metre and is equal to 2.113433 pints.

The gramme or unit of weight is the weight of that quantity of distilled water at its maximum density, 4° C., which will fill the cube of $\frac{1}{100}$ part of a metre, and is equal to 15.4323 + grains.

In obtaining the multiples and subdivisions of all these units the number ten is used exclusively. The latin prefixes deci, centi and milli, meaning $\frac{1}{10}$ $\frac{1}{100}$ and $\frac{1}{1000}$ are used in connection with the name of each unit to denote its subdivisions, while the Greek prefixes, Deka, Hecto and Kilo, meaning 10, 100 and 1000 are used in the same way to denote multiplication of the unit. Thus one decigramme $= \frac{1}{10}$ gramme, one centigramme $= \frac{1}{100}$ gramme, and one milligramme $= \frac{1}{1000}$ gramme, and in the same way one decigramme $= 10$ grammes one hectogramme $= 100$ grammes and one kilogramme $= 1000$ grammes.

Table of metric system of weights and equivalents in grains.

One Milligramme	.0154 grains.
One Centigramme....	.1543 grains.
One Decigramme.....	1.5434 grains.
One Gramme........	15.432 grains.
One Decigramme	154.3234 grains.
One Hectogramme ...	1543.2348 grains.
One Kilogramme	15432.3487 grains.

Table of Metric System of Capacity and equivalents in Apothecaries' measure:

Kilo-litre.................	264.19 gallons.
Hecto-litre	26.419 gallons.
Deci-litre	2.6419 gallons.
Litre.....:..............	2.1135 pints.
Deci-litre.............	3.3816 fluid ounces.
Centi-litre..........	2.7053 fluid drachms.
Milli-litre	16.2318 minims.

SYNONYMOUS NAMES.

Aqua Saturni, Liq. Plumbi Subacet.
Aqua Sedativa, Sedative Water.
Aqua Javelle, Liquor Potassæ Chloratae.
Aqua Luciæ, A kind of liquid soap.
Aqua Phagedoenica Flava, Lotio Flava.
Aqua Phagedoenica Nigra, Lotio Nigra.
Acidum Phenicum, Carbolic Acid.
Acetum Saturni, Liquor Plumbi Subacetatis.
Acetum Plumbi, Liquor Plumbi Subacetatis.
Aquila Alba, Calomel
Balsamum Traumaticum, ⎫
Balsamum Friar's,   ⎪   A compound Tr. of
Balsamum Jesuit,   ⎬   Benzoin.
Balsam de Maltha,   ⎭
Basham's Mixture, Liq. Ammo Acetatis.
Boultons Solution, Liquor Iodi Carbolatus.
Bark, A term applied to the different species of
    Cinchona.
Bland's Pills, Pilulæ Ferri Carbonatis.
Bitter Salz, Magnesia Sulphas.
Blutwerzel, Bloodroot.
Carbasus Iodoformata, Iodoform Gauze.

Crocus Martis, Oxide of Iron.
Carron Oil, Lime Liniment.
Calomelas, Calomel.
Decoctum Zitmanni, A compound decoction of
  Sarsaparilla.
Elixir Curassao, Elixir Curacao.
Extract Goulard, Liquor Plumbi Subacetatis.
Fleming's Tinct. of Aconite, Aconite root 10 oz
  Troy Alcohol Q. S. 15 f oz.
Flores Benzoes, Benzoic Acid.
Ferri Quevenne's, Reduced Iron.
Gossypium Stypticum, Stiptic Cotton.
Glonoin, Nitro-glycerine.
Goulard's Extract, Liquor Plumbi Subacetatis.
Gummi Mimosæ, Gum Acacia·
Gutta Bateman, Bateman's Pectoral Drops.
Gutta Nigra, Acetum Opii.
Heira Picra, Pulvis Aloes et Canellæ.
Iodine Caustic, Liquor Iodi Causticus.
James' Powder, Pulvis Antimonialis.
Jesuits Balsam, Tr. Benzoin Comp.
Kali or Kalium, Potassa
Lac Fermentatum, Kumyss.
Liquor Arsenicalis, Fowler's Solution.
Labarraque's Solution, Liquor Sodæ Chloratae.
Lapis infernalis, Nitrate of Silver.
Mistura Gummosa, Mistura Acaciæ.
  "   Basham, Liq. Ammo. Acetatis.
Mercurius, Mercury.
Mercurius Bismuithi, Subnitrate of Bismuth.
Number Six, Tinctura Capsici et Myrrhæ.
Natri or Natrium, Soda.
Natro, Kali acidulum tartaricum, Rochelle Salts.

Nihilum Album, Oxide of Zinc.
Oleum Anthos, Oil Rosemary.
Oleum Harlemensis, Harlem Oil.
" De Cedro, Oil of Lemon.
" Carron, Lime Liniment.
" Waldwoll, Oil of pinus pumilis.
Opodeldoc, Linimentum Saponis Camphorata.
Opodeldoc, Is often written when Linimentum
Saponis is wanted.
Pilulæ Ad Prandium, Dinner Pills.
Pulvis Kurellæ, Comp. Licorice powder.
Quevenne's Iron, Reduced Iron.
Solution Labarraque, Liquor Sodæ Chloratæ.
Solution Donovan, Liquor Arsenii et Hydrargyri
Iodidi.
Solution Monsel's, Liquor Ferri Tersulphatis.
Solution Boulton, Liquor Iodi Carbolatus.
Solution Vleminck's, Liquor Calcis Sulphuratæ.
Solution Lugols, Liquor Iodi Compositus.
Solution Fowler's, Liquor Potassii Arsenitis.
Solution Villate's, Mistura Adstringens et Es-
charotica.
Sal Monsel, Persulphate of Iron.
Sal Amarum, Epsom Salts.
Sal Mirabile, Plumbe's Salts.
Spiritus Glonoini, a one per cent. solution of
nitroglycerine in alcohol.
Spiritus Mindererus, Liquor Ammonii Acetatis.
Spiritus, Alcohol.
Species Laxantes, St. Germain Tea.
" Pectorales, Breast Tea.
Syrupus Doveris, Syrupus Ipecacuanhæ et Opii.
Thebaica, Opium.

Tinctura Warburg, Tinctura Antiperiodica N. F.
Tinctura Pectoralis, Pectoral Tincture N. F.
"   Huxhams, Compound Tincture of Cinchona.
Tincture Thebaica, Tinc. Opium.
"   Ferri Pomata, Tincture of Ferrated Ext. of apples.

Theriaca ⎱
Treacle   ⎰ Molasses.

Unguentum Matris, Mother's Salve.
Zinci Flores, Oxide of Zinc.

INDEX.

Abbreviations............................ 72
Capsules 59
Compounding Prescriptions............... 14
Compounding Liquid Preparations.......... 22
Domestic Measures....................... 80
Emulsions............................... 41
Explosive Prescriptions.................. 69
Filing Prescriptions..................... 64
Incompatibility.......................... 15
Metric Prescriptions..................... 21
Pills.................................... 45
Powders................................. 62
Reading Prescriptions.................... 11
Solvents................................ 76
Suppositories........................... 54
Symbols 20
Synonyms................................ 83
The Prescription........................ 5
The Prescription Case................... 66
Toxicology.............................. 77
Weights and Measures.................... 80
Writing Prescriptions................... 7

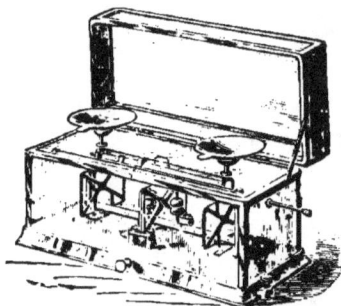

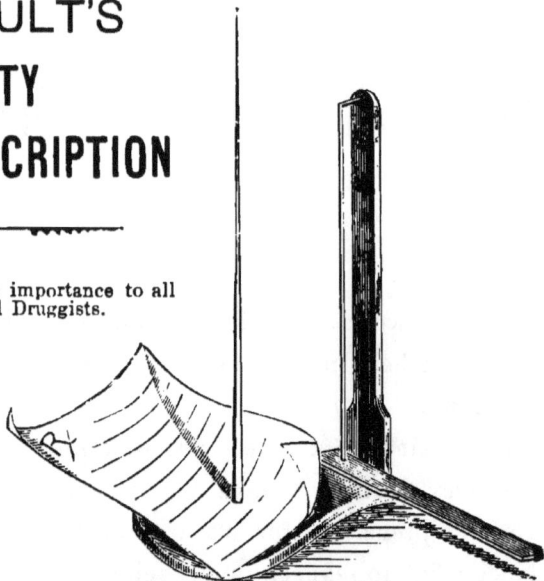

Gray's Elements of Pharmacy.

~୧ଚ⊚⊚⊙ଚ౨~

A short, practical course in the rudiments of Pharmacy, designed for the use of those just beginning the study. The course of study in Pharmacy, Botany and Chemistry, which is mapped out and treated of in this work, is such as will enable the student to make the most rapid progress from the beginning. Neatly bound in cloth. Price $1.50.

~୧ଚ⊚⊚⊙ଚ౨~

For Sale by

GRAY & BRYAN, Publishers,

BOX 593, CHICAGO, ILL.

HANCE BROS. & WHITE,

Pharmaceutical Chemists,

PHILADELPHIA: Callowhill and Marshall Streets.

NEW YORK: 17 Platt St.   BOSTON: 11 Portland St.

CHICAGO: 59 Lake St.   PITTSBURGH: Bissell Building.

Our Red Messina Orange has a way of bringing people back for more; and people don't come back for more without talking about your soda.

Nobody knows the value of soda except two sorts of people: First, anybody on a hot day; second, druggists that serve it in just four ways: Onethly, cool; twothly, in thin glasses; threethly, with cheerful courtesy; fourthly, with our fruit juices.

The card to hang on your fountain is 7x11, perfectly plain. All the ornament on it is a silk cord to hang it by. The signs that bring folks in are plain and easy to read, and true. The easy and pleasant reading brings a customer once. When he finds it true, he comes again. This coming again is what we think most of. Red Messina Orange does it. Soda fountains are doubling their business in it. But serve it in those four ways. Not even Red Messina Orange is rich and fine and inspiring enough to be served without one of the four accompaniments: Coolness, daintiness, courtesy—what was the fourth? Why, the pirates are imitating Red Messina Orange already—look out.

Order a dozen or two and use a bottle; serve it right, and return the rest if it does not hit the bull's eye.

It was new last year; it is new this year to nine out of ten. Put it forward. The people have no means of finding it out, and every man and woman will thank you for introducing it.

That's why you ought to have our card at your fountain. It is better than a sign that would cost you a dollar to get up.

If you have not received it, or if the one you have is not fresh and clean, we will send you a new one free—two or three if you like—on application.

—— GET ——

GRAY'S URINALYSIS

· · · FOR · · ·

COMPLETE AND IMPROVED DIRECTIONS

FOR TESTING URINE.

———

Remembering that the Pharmacist is very often called upon by the Physician to test suspected samples of urine, we have tried in this pamphlet to give the best and simplest processes for procedure.

———

PRICE BY MAIL, POSTAGE PREPAID, 25c.

—

· · · ORDER OF · · ·

GRAY & BRYAN. Publishers,

Box 593, CHICAGO, ILL.